# HOW TO READ A WEATHER MAP

Black lines on a weather map represent isobars. The prefix *iso* means "equal" and bars or millibars are the measure of atmospheric pressure. An isobar is therefore a line of equal pressure. Lower numbers represent lower pressure and higher numbers reflect higher pressure.

The blue line with triangles on it is a cold front, with the triangles pointing in the direction in which the front is moving. The red line with half circles is a warm front, with those half circles pointing in the direction in which the front is moving. Note that the fronts lie in areas where the otherwise circular rings around the center of low pressure are contorted. Other distensions in these concentric circles represent troughs and are likely areas of precipitation.

Though not used in this example, other common features on surface maps are stationary fronts, represented by an alternating line of blue triangles and red half circles pointed in opposite directions; occluded fronts, which are purple lines with half circles and triangles all pointing in the same direction; and brown or purple dashed lines, which are catchall symbols, either for surface troughs or dry lines.

The heat index approximates what the temperature will feel like with varying combinations of heat and relative humidity, as well as indicates the danger of particular combinations. If temperatures are high enough, it doesn't take much to turn a warm day into a potentially dangerous one.

**Heat Index**

| | | | | | | | | TEMPERATURE (°F) | | | | | | | | |
|---|---|---|---|---|---|---|---|---|---|---|---|---|---|---|---|---|
| | | 80 | 82 | 84 | 86 | 88 | 90 | 92 | 94 | 96 | 98 | 100 | 102 | 104 | 106 | 108 | 110 |
| RELATIVE HUMIDITY (%) | 40 | 80 | 81 | 83 | 85 | 88 | 91 | 94 | 97 | 101 | 105 | 109 | 114 | 119 | 124 | 130 | 136 |
| | 45 | 80 | 82 | 84 | 87 | 89 | 93 | 96 | 100 | 104 | 109 | 114 | 119 | 124 | 130 | 137 | |
| | 50 | 81 | 83 | 85 | 88 | 91 | 95 | 99 | 103 | 108 | 113 | 118 | 124 | 131 | 137 | | |
| | 55 | 81 | 84 | 86 | 89 | 93 | 97 | 101 | 106 | 112 | 117 | 124 | 130 | 137 | | | |
| | 60 | 82 | 84 | 88 | 91 | 95 | 100 | 105 | 110 | 116 | 123 | 129 | 137 | | | | |
| | 65 | 82 | 85 | 89 | 93 | 98 | 103 | 108 | 114 | 121 | 128 | 136 | | | | | |
| | 70 | 83 | 86 | 90 | 95 | 100 | 105 | 112 | 119 | 126 | 134 | | | | | | |
| | 75 | 84 | 88 | 92 | 97 | 103 | 109 | 116 | 124 | 132 | | | | | | | |
| | 80 | 84 | 89 | 94 | 100 | 103 | 113 | 121 | 129 | | | | | | | | |
| | 85 | 85 | 90 | 96 | 102 | 110 | 117 | 126 | 135 | | | | | | | | |
| | 90 | 85 | 91 | 98 | 105 | 113 | 122 | 131 | | | | | | | | | |
| | 95 | 86 | 93 | 100 | 108 | 117 | 127 | | | | | | | | | | |
| | 100 | 87 | 95 | 103 | 112 | 121 | 134 | | | | | | | | | | |

**Likelihood of Heat Disorders with Prolonged Exposure or Strenuous Activity**

Caution    Extreme Caution    Danger    Extreme Danger

*While we often think of them as a simple background to our daily lives, cloud types can tell you a lot about the weather.*

## CLOUD BASICS

- Lower clouds are more likely to produce precipitation.
- Cooler temperatures generally lead to lower cloud bases.
- More bulbous, textured clouds indicate that there's more activity in the atmosphere; such clouds are more likely in warm conditions.

**Cirrus** are high, wispy clouds that are made of ice crystals. Generally speaking, they are fair-weather clouds. If cirrus clouds are accumulating, they may be at the leading edge of an advancing storm system, though there would be no immediate threat.

**Cirrostratus** are a blanket of high-level clouds. They may precede a larger storm system, but there would rarely be an immediate threat.

3

**Cirrocumulus (herringbone)** patterns in clouds generally mean that winds at cloud level are brisk. When they appear in high clouds, it may correlate with the jet stream. The "ribs" of the cloud will be perpendicular to the wind. If chevrons form, they will point in the direction that the wind is blowing, and will suggest the direction that any inclement weather will move.

**Altocumulus** are mid-level clouds that rarely portend bad weather, though they may, in isolated circumstances, produce a couple of drops of rain. They generally appear as a broken sheet of clouds across the sky, often in a pattern, such as "puffs" or striations.

**Altostratus** are sheets of somewhat thin clouds that are higher in the sky than stratus, but thicker and lower than cirrostratus. They tend to be gray, but not as dark gray as nimbostratus clouds. They often appear in conjunction with warm fronts, and are replaced by nimbostratus as rain approaches. They easily reflect wind patterns, producing ripples in an otherwise flat texture, or developing gaps.

**Stratocumulus** clouds blanket the sky with a patchwork of clouds. Generally speaking, stratocumulus are not a source of a lot of rain, though some light rain is possible with them. If they are generally moving in from the west, they can indicate thicker clouds or precipitation that is on the way.

**Stratus** clouds give us "overcast" skies. While stratus may produce precipitation, lower, darker clouds (known as nimbostratus clouds), generally carry a better chance of rain or snow.

**Cumulus** are bulbous, puffy clouds that are most common in warmer temperatures. Small isolated cumulus clouds are common in fair weather, but large and growing clouds, especially when joined by other similar clouds, are indicative of developing rain or thunderstorms.

**Cumulonimbus** are mountainous, bulbous clouds that give way to heavy precipitation, and quite often thunder and lightning, along with the other hazards of thunderstorms. Cumulonimbus clouds are generally so vast that if you can see the top of them, they are quite a distance away. As they approach, you will only see the base of the clouds, which may give rise to other structures, like wall clouds, shelf clouds, or mammatus.

**Anvil** clouds are a subset of cumulonimbus clouds that form when rising updrafts reach the tropopause (the boundary with the stratosphere) and flatten, causing the cloud to resemble an anvil. Sometimes, the clouds bubble beyond the tropopause thanks to the momentum. This is called an "overshooting top." Anvil clouds and their violent updrafts are most likely to produce heavy rain and hail.

**Mammatus** are lobe-shaped clouds that are formed by downdrafts. Mammatus are generally a sign that thunderstorms are collapsing, and are often seen at the tail end of strong storms.

**Shelf** clouds form at the leading edge of advancing storms, such as during strong straight-line wind events. Unlike wall clouds, shelf clouds can extend along the entire base of the cloud, and move rapidly in the direction of storm movement. They are usually accompanied by extremely strong winds, but are not typically tornadic in nature.

**Wall** clouds descend from thunderstorms, and are often seen on the southwestern flank of thunderstorms. The clouds rotate and may produce funnels or tornadoes.

**Funnel** clouds consist of a rotating column of wind and descend from wall clouds where they pull in moisture at the base of the wall cloud. A tornado exists when the rotating air reaches the ground. Tornadic winds are not always visible, and if a funnel is noted, it should be treated as a tornado (i.e., seek shelter).

**Roll** clouds are similar in meaning and appearance to shelf clouds; however, they are often detached from the greater storm structure.

**Fractus** are stray bits of cloud that have broken away from larger masses. They can often be confused with funnels or tornadoes when they appear below stratus or thunderstorm clouds.

**Contrails** form when an aircraft moves through a nearly saturated bit of air, and water condenses in the turbulent flow, forming trails in the sky.

**Noctilucent** clouds are extremely high-altitude clouds that are generally invisible until twilight; they are most often seen at high latitudes.

**Fallstreaks** are circular gaps in a blanket-like layer of clouds such as altostratus. They are formed by an area of super-cooled cloud droplets freezing and falling, leaving a hole in the cloud.

**Lenticular** clouds are often found near and above mountains, and they can often look like flying saucers. They are formed as air tries to move around the mountains.

**Asperitas** cloud formations are known for wavy bases and can be associated with precipitation, though they aren't indicative of major storms.

**Gravity waves** produce rolls of clouds; they occur as gravity works to achieve atmospheric equilibrium. In an unstable atmosphere, gravity waves may lead to rain or isolated thunderstorms.

**Fog** typically develops overnight, but also frequently after heavy rain. It's an indication of low-level moisture and cooler temperatures. It generally dissipates with warming temperatures.

## DIFFERENT-HUED SKIES

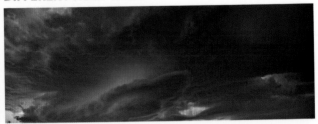

**Dark blue** (relative to the time of day) **and green** skies are evidence of moisture in the sky that can blot out light. At the very least they suggest heavy rain, but may mean hail or suggest a strong storm.

**Yellow** skies occur when ice crystals diffuse light, and can mean snow or sleet. Yellow skies can also be caused by smoke lingering high in the atmosphere from distant fires.

**Red and pink** skies are seen at sunrise or sunset, and are a result of the low sun angle and refraction through moisture in the area. Red-tinted skies at sunrise may be indicative of pending rain during the day.

*In a heavy rain it may only take minutes to produce flooding.*

## PRECIPITATION BASICS

- Darker clouds bear more water and often lead to heavier precipitation.
- Storms generally move from the west to east, and less frequently from the south or north. They almost never move from the east.
- Stronger storms lead to larger raindrops, and large raindrops may portend hail.
- Snow falls in clumps when it's warmer outside, but if the air is very cold, individual flakes get larger because of the crystallization process.

**Mist**

**Mist:** Drops that are small enough to remain suspended in air, but heavy enough to condense on objects.

**Drizzle:** Falling liquid precipitation that's less than .5 mm in diameter.

**Rain:** Falling liquid precipitation that's greater than .5 mm in diameter.

**Freezing drizzle:** Supercooled liquid droplets that are less than .5 mm in diameter and freeze on contact with a surface.

**Freezing rain:** Supercooled liquid droplets that are greater than .5 mm in diameter and freeze on contact with a surface.

**Sleet:** Supercooled droplets that freeze shortly before reaching the ground and fall as pellets.

**Graupel:** Also known as soft hail, it forms when supercooled droplets freeze as snow, and fall as clumps. It is usually larger than sleet.

**Snow:** Frozen precipitation in the form of crystals. It becomes more crystalline with colder temperatures, and tends to fall more gently. When it's warmer and closer to freezing, it can fall as heavily as rain.

**Graupel**

**Hail:** Frozen precipitation during warm weather. Hail occurs when ice forms high in thunderstorm clouds. The hailstones grow by accumulating moisture, which freezes on their surfaces. The stone falls when it is too heavy to be supported by thunderstorm updrafts.

**Hail**

*Fair-weather phenomena can be as fascinating as cloud types.*

### 22-degree Halo

**Halos** are rings of light that surround the sun, and sometimes, the moon. They occur when ice crystals refract light around the sun in a circle, or arcs. They often occur in conjunction with high cirrus clouds, and usually appear in cold weather. The refracted light can appear in a variety of arrangements. The 22-degree halo appears at a 22-degree radius from the sun. A panhelic circle runs perpendicular to the halo, while an arc tangent is a partial mirror to the halo. They all are caused by a refraction of light by ice crystals, with the phenomena more likely at colder temperatures.

### Panhelic Circle

### Upper Arc Tangent

**Sun Pillars** are similar to halos, in that they are caused by refraction from ice crystals, but they are vertical shafts of light extending from the sun. They occur when the sun is on or near the horizon.

**Sun Dogs** are similar to sun pillars and appear as bright points of light parallel to the sun (or moon, in the case of moon dogs), and lack some of the vertical dimension pillars have.

**Rainbows** are low-level refractions of light that appear after rain. Suspended drops act as prisms breaking the white sunlight into its component colors. Rainbows are only visible with the sun at your back.

A **corona** is a ring around a circular or spherical object. Meteorologically, it is represented by a translucent disk around the sun or moon. They are caused by thin clouds, often at mid or lower levels, scattering out the light from the source.

**Aurora borealis/northern lights** are the visible chemical reaction caused by charged particles from the sun reacting with gases high in the Earth's atmosphere. They are usually seen at high latitudes, though on rare occasions they can be seen much farther south.

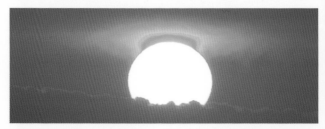

**The green flash** is a unique phenomenon caused by sunlight refracting within the Earth's atmosphere. It occurs most frequently as the sun disappears over the horizon at sunset. The best place to see the green flash is along the West Coast, or the Gulf Coast of Florida.

**Heat lightning** is a misnomer; it's often used to describe distant lightning that is not associated with local thunder. In reality, it's usually lightning occurring at the top of a very distant thunderstorm, potentially up to 100 miles away, depending on the viewer's vantage point.

**Pyrocumulus** clouds are formed by smoke and the rapid updrafts caused by intense heating at the surface. They can either be caused by wildfires or volcanic eruptions, and can function like cumulonimbus, producing lightning and even small tornadoes. Pyrocumulus are best observed from a great distance.

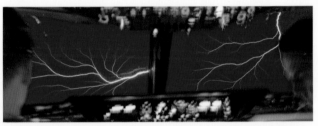

**St. Elmo's Fire** is a phenomenon sometimes seen in thunderstorm conditions and even aboard aircraft. An electrified glow extends from spires or sharp points (such as aircraft wings or ship masts) usually toward the storm. It appears as a bluish haze and is caused by the ionization of particles around a conductor.

*To stay safe, make sure you know the weather forecast where you are.*

The media is one of the greatest tools for ensuring your weather safety. Broadcast television stations relay National Weather Service weather watches and warnings, particularly for severe thunderstorms, tornadoes, and flash floods. AM Radio is also a crucial link when it comes to providing the public with severe weather updates. What's more, in weather emergencies, such as tornadoes in metropolitan areas, the Emergency Alert System can override broadcasts on TV or radio, and even send messages to cell phones to encourage residents to seek shelter, evacuate, or make other preparations for dangerous weather.

## WATCHES AND WARNINGS APPLY TO SPECIFIC GEOGRAPHIC AREAS

Be mindful that watches, warnings, and advisories are almost always issued with reference to counties (or parishes, depending on state jurisdictions), so it is important to know the name of the county you find yourself in. In states with many different counties—Minnesota has 87—this isn't always apparent if you're traveling, so be alert to the names of counties you are traveling through if weather conditions are not good.

## WEATHER WATCHES

After you have familiarized yourself with your local geography, you need to know the difference between a weather **watch** and a weather **warning**.

A weather watch, such as one for a severe thunderstorm or a tornado, means that conditions are favorable for the development of the extreme weather in question. For example, if you're in an area with a severe thunderstorm watch, a severe thunderstorm may not be present in the area, but conditions are right for one to form. As a result, a fairly large area is being monitored for an extended period of time for the development of those storms.

### A Weather Watch Example

A typical weather watch, say one issued at 2:00 p.m. in Minnesota, would say something like: "A Tornado Watch is in effect within 100 NM (nautical miles) of a line from 20 miles northwest of St. Cloud to 15 miles west of Worthington until 11:00 p.m. This Tornado Watch is in effect for the following counties…"

The counties affected would then be listed off in alphabetical order, as would the threats that tornadoes pose. With a weather watch, it is important to stay tuned for weather events, but there is no imminent or immediate danger.

## WEATHER WARNINGS

When there is a weather warning, it means that severe weather is occurring and expected to arrive soon in the warned location, if it is not there already. Weather warnings are focused on a smaller area and last for a shorter amount of time.

### A Weather Warning Example

Again, an example is helpful. A tornado warning would read something like the following: "A tornado warning has been issued for Kandiyohi County, Minnesota and western Meeker County, Minnesota. At 4:42 p.m., Doppler Radar indicated a tornado 3 miles east of Raymond, or 10 miles southwest of Willmar. This possible tornado is moving east northeast at 30 mph. It is expected to be near Svea at 4:46, Rosendale at 4:52, Litchfield at 4:59. This warning is in effect until 5:25 p.m." The warning would then list the threats posed by tornadoes, and advise residents on the best course of action.

*If severe weather strikes, seek shelter right away.*

## THUNDER AND LIGHTNING

Cloud-to-ground lightning takes the easiest path to the earth, and that is often by striking the tallest object. So if you hear thunder, it's important to not be the tallest object in your area. Also, don't stand near or under other objects that could be struck, because lightning bolts are capable of arcing from that object to you on their way to the ground. Trees are good conductors, and they therefore provide poor shelter from lightning.

If you are unable to get to safety and are in an open area, lightning-strike survivors say that you can feel a rise in static electricity shortly before lightning strikes. If you have any reason to believe that a lightning strike is possible, the safest thing to do in an open field is to crouch down and give yourself a low profile.

## TORNADOES

When it comes to storms with strong winds and/or tornadoes, the best thing is to seek shelter indoors. Basement stairwells or storm cellars, well away from windows, are ideal. If you're caught outside, the best thing to do is keep a low profile. Tall vehicles like trucks or vans can be blown over, and remaining upright exposes you to more debris. Stay low and out of the wind if you are unable to find shelter.

One of the most dangerous things that people do when tornadoes approach is to take shelter in a bridge abutment or underpass. The constricted area of an underpass creates a Bernoullian flow. That is to say, air moves through smaller areas faster. The underpass will cause winds to be even stronger, and moving up the side of the bridge abutment will only compound the problem.

## HEAVY WINDS

When it comes to heavy winds and hail—whatever the cause—stay indoors. When winds are strong, airborne debris are a real danger, so you need strong walls and a roof overhead.

## FLOODING

When an area is flooded, it's impossible to tell how deep the water is just by looking at it, and it's also impossible to ascertain how strong its current is. That's why it's always important to stay out of floodwaters, especially if you're in a vehicle. It only takes a few inches of water to move a car, and many flooding deaths occur when drivers try to pass through a flooded area only to discover that it's deeper than their vehicle is tall. Floods often wash out roadways, too.

Flooding can often necessitate evacuations. Seasonal river flooding leads to evacuations near the river's banks, and they tend to give affected residents a little bit of time to pack and move important belongings to higher levels. Flash flood-induced evacuations are much more urgent, and often indicate mudslides or a levy or dam failure, and they could signal catastrophe on the way. If a flash flood evacuation is issued in your area, get out immediately.

## HIGH TEMPERATURES

When temperatures are high, stay hydrated if you must remain outdoors, and make sure that you and your loved ones have adequate access to air conditioning. Hyperthermia and heat-related illnesses don't just occur in cases of physical exertion; many fatal cases occur in people who are simply stuck indoors without a way to cool down.

## WINTER TRAVEL

The snow and ice layered on a road melt ever so slightly when a car travels over them, creating a greasy, slick surface that makes driving difficult. In particularly cold weather, black ice, which is invisible to the naked eye, can form in the cracks and crevices of the payment.

If the conditions are treacherous, there isn't much that anyone can do except drive responsibly and slowly. Don't put yourself in a position where you have to quickly brake or accelerate or change direction. When you do need to brake, do so gradually.

Should things go wrong and you get stranded in your vehicle, it's important to have a safety kit, including blankets, flares, food, and perhaps a shovel, or sand for traction if you are stuck in snow.

*By paying attention to your surroundings, you can get a rough idea on what weather is in store.*

**1.** Assuming there isn't obviously a storm bearing down on you (dark clouds, increasing winds, etc.) identify the direction the wind is coming from. Stand with your back to the wind, and point straight to the left. In doing so, you are pointing in the direction of low pressure. If you find yourself pointing to the west, northwest, or slightly to the southwest, the threat of rain is increasing.

**2.** Dark clouds are an obvious clue that inclement weather is approaching, but that simple fact bears repeating. In the summer, that is a sure sign that those clouds are laden with moisture. If you note that the clouds are dark, and the wind is blowing toward the storm, be careful, because that suggests localized low pressure, strong updrafts, and the threat for hail or even tornadoes in conjunction with the already expected heavy rain. If wind is blowing from the storm toward you, the storm is likely collapsing, and the "outflow" is caused by the weight of falling rain suppressing the updraft, and dispersing all that energy. The exception is when storms

23

are strong and linear. In those cases, very strong winds will precede the advance of the thunderstorm, in the direction of the storm's motion, though wind is likely to be drawn into the storm before it turns around. Be wary of the so-called "calm before the storm" which is likely to precede the leading edge of thunderstorm-generated winds.

When observing clouds, keep in mind that clouds shift and change all the time. The most informative time for monitoring cloud changes is when you can see the cloud tops. If they are growing and building, be prepared for some nasty weather. The taller they get, the more likely that hail is in the future. Generally speaking, rapid cloud growth is indicative of instability in the atmosphere, and that is a sign that thunderstorm activity could erupt rapidly.